SIMPLE MACHINES ALL AROUND US

# INCLINED PLANES AT WORK

I0795312

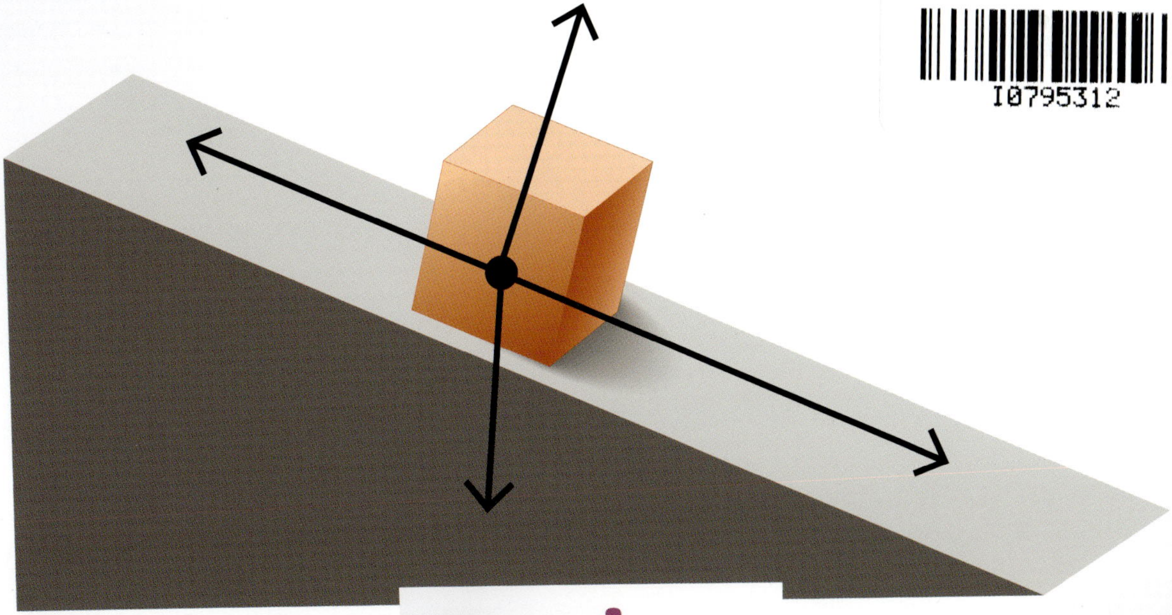

BY MAGGIE S. WESLEY

Enslow PUBLISHING

DISCOVER!

**Please visit our website, www.enslow.com. For a free color catalog of all our high-quality books, call toll free 1-800-398-2504 or fax 1-877-980-4454.**

**Library of Congress Cataloging-in-Publication Data**
Names: Wesley, Maggie S., author.
Title: Inclined planes at work / Maggie S. Wesley.
Description: New York : Enslow Publishing, 2022. | Series: Simple machines all around us | Includes bibliographical references and index.
Identifiers: LCCN 2020013402 | ISBN 9781978520806 (library binding) | ISBN 9781978520783 (paperback) | ISBN 9781978520790 (set) | ISBN 9781978520813 (ebook)
Subjects: LCSH: Inclined planes–Juvenile literature.
Classification: LCC TJ1428 .W47 2022 | DDC 621.8–dc23
LC record available at https://lccn.loc.gov/2020013402

Published in 2022 by
**Enslow Publishing**
101 West 23rd Street, Suite #240
New York, NY 10011

Copyright © 2022 Enslow Publishing

Designer: Katelyn E. Reynolds
Editor: Kristen Nelson

Photo credits: Cover, pp. 1 Designua/Shutterstock.com; cover, pp. 1–24 (background and typeface) Reytr/Shutterstock.com; cover, pp. 1–24 (paper) jannoon028/Shutterstock.com; cover, pp. 1–24 (note sticker) Kindlena/Shutterstock.com; p. 5 abalcazar/E+/Getty Images; p. 7 Boris SV/Moment/Getty Images; p. 9 FooTToo/Shutterstock.com; p. 11 Mitch Diamond/ The Image Bank Unreleased/ Getty Images; p. 12 Tom Grubbe/Moment/Getty Images; p. 13 Vicki Jauron, Babylon and Beyond Photography/Moment/Getty Images; p. 14 Kitti Boonnitrod/Moment/Getty Images; p. 15 (Galileo) Stock Montage/Getty Images; p. 15 Dorling Kindersley/Getty Images; p. 16 Xia Yang/Moment/Getty Images; p. 17 RyanJLane/E+/Getty Images; p. 19 Andrea Nissotti / EyeEm/Getty Images; p. 21 (ramp) BRIAN MITCHELL/ Corbis Documentary/Getty Images; p. 21 (driveway) William Manning/ Corbis Documentary / Getty Images Plus; p. 21 (screws) Alexandrum79/ iStock / Getty Images Plus; p. 21 (pipes) DIY Photolibrary/Construction Photography/ Avalon/Getty Images; p. 21 (stairs) Marizza/E+/Getty Images; p. 21 (slides) Sergei Kozak/ Corbis / Getty Images Plus.

Portions of this work were originally authored by Gillian Houghton Gosman and published as *Inclined Planes in Action*. All new material this edition authored by Maggie S. Wesley.

All rights reserved. No part of this book may be reproduced in any form without permission in writing from the publisher, except by a reviewer.

Printed in the United States of America

Some of the images in this book illustrate individuals who are models. The depictions do not imply actual situations or events.

CPSIA compliance information: Batch #CSENS22: For further information contact Enslow Publishing, New York, New York, at 1-800-398-2504.

Find us on

# Contents



**Boldface** words appear in Words to Know.

# The Incredible Inclined Plane

It may not look like it, but a **ramp** is a special kind of machine called an inclined plane. "Inclined" means one end is raised. Inclined planes are simple machines. These are tools that change **motion** and certain forces to make work easier.

LIKE THE FIVE OTHER SIMPLE MACHINES, INCLINED PLANES DON'T HAVE MANY PARTS.

# What Makes an Inclined Plane?

Inclined planes are a surface, or upper part of something, that is sloped, or has a downward **pitch**. Objects can slide up or down this surface. The steeper the slope, the more energy, or power, is needed to move the object up the slope.

THE STEEPER AN INCLINED PLANE'S SURFACE, THE FASTER OBJECTS CAN MOVE DOWN IT.

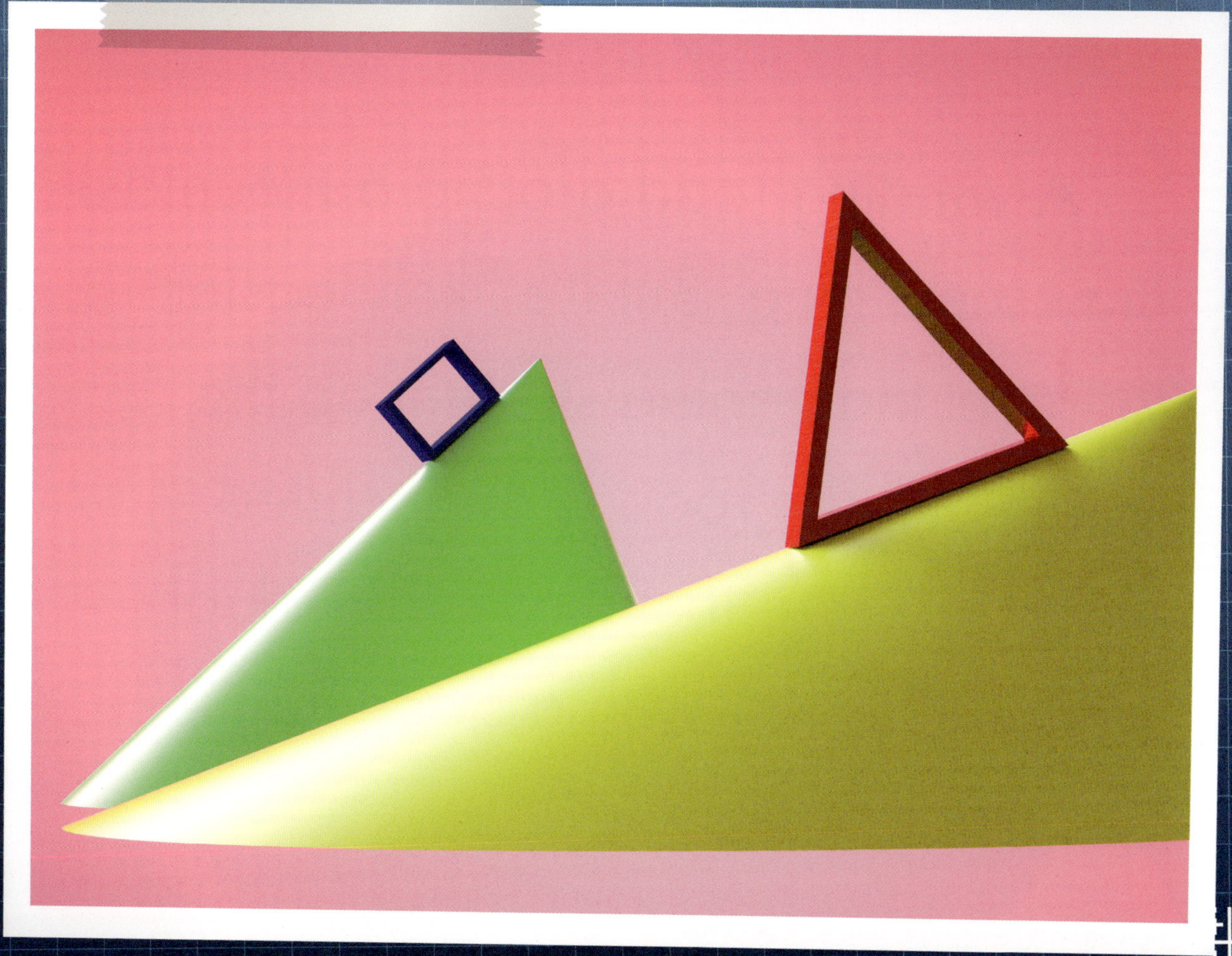

# The Long Way's Advantage

A winding road around a mountain may seem long. But it's likely easier to travel than going straight up the mountain! The road is an inclined plane because it makes the work of going up easier with a slope that is less steep.

AN INCLINED PLANE'S MECHANICAL ADVANTAGE IS THE ADVANTAGE, OR BENEFIT, GAINED BY USING A MACHINE OR TOOL TO CHANGE A FORCE.

# Figuring Out the Forces

An inclined plane's **angle** decides how much easier it makes work. A narrower angle means a less steep slope and less force needed to push objects up it. A wider angle means a steeper slope, and more force is needed.

A WIDER ANGLE ALSO MEANS A STEEPER SLOPE AND FASTER SPEEDS FOR OBJECTS THAT GO DOWN INCLINED PLANES—LIKE ROLLER COASTER CARS!

# Who Invented the Inclined Plane?

A person didn't invent the inclined plane. Nature did! A hill is an example of an inclined plane found in nature. However, people have invented special inclined planes to use in their daily lives, like stairs. They've also used inclined planes in **compound machines**.

WILD ANIMALS LIKE BUFFALO USE AN INCLINED PLANE WHEN THEY WALK UP A HILL.

## The History of Inclined Planes

The towering **pyramids** in Egypt wouldn't exist without the inclined plane! People pushed big blocks up ramps to add them to the pyramids. Italian scientist Galileo Galilei used an inclined plane around 1604 to make important discoveries about motion.

GALILEO FIGURED OUT THAT AN INCLINED PLANE'S DOWNWARD SLOPE KEEPS A BALL IN MOTION.

# Inclined Planes on the Job

When you slide down a slide, you are using an inclined plane. However, there are many more uses for doing work. Driveways that are higher at one end and staircases are inclined planes that move things—such as people—from a lower place to a higher place.

A SLIDE IS AN EXAMPLE OF A CHUTE, OR AN INCLINED PLANE THAT THINGS CAN SLIDE DOWN.

# When Tasks Get Trickier

Inclined planes don't just move things up and down. An inclined plane wrapped around a **cylinder** makes another simple machine—the screw. An inclined plane also creates a zipper's slide. Another simple machine called a wedge makes the zipper's teeth.

A SCREW IS OFTEN USED TO CONNECT TWO OBJECTS.

# Find Inclined Planes Around You

Now that you know what inclined planes are, can you spot any around you? They might even be hiding in the walls as **diagonal** pipes! Even when we can't see them, inclined planes and other simple machines continue to make our lives better.

RAMPS HELP PEOPLE WHO USE WHEELCHAIRS ENTER BUILDINGS AND OTHER PLACES.

RAMP

DRIVEWAY

SCREW

PIPES

STAIRS

SLIDE

# Words to Know

**angle** The space between two lines or surfaces that come together.

**compound machine** A tool or machine that uses more than one simple machine.

**cylinder** A shape that has two circular ends and straight sides.

**diagonal** Slanting from one side to another.

**motion** Movement.

**pitch** The amount of slope a surface has.

**pyramid** A 3-D shape that has a rectangular base and triangular sides.

**ramp** An object that connects two surfaces that are at different levels.

# For More Information

## Books

Marsico, Katie. *Simple Machines.* Ann Arbor, MI: Cherry Lake Publishing, 2019.

Rivera, Andrea. *Inclined Planes.* Minneapolis, MN: Abdo Zoom, 2017.

Troupe, Thomas Kingsley. *Keep It Simple, Rapunzel!: The Fairy-tale Physics of Simple Machines.* North Mankato, MN: Picture Window Books, 2019.

## Websites

**Physics for Kids: Simple Machines**
*www.ducksters.com/science/simple_machines.php*
The Ducksters crew helps you learn more about simple machines on this site!

**Simple Machines | Science Trek**
*ny.pbslearningmedia.org/resource/idptv11.sci.phys.maf.d4ksim/simple-machines/*
Check out a cool video and learn more about the simple machines in your daily life!

**What Is a Simple Machine?**
*www.wonderopolis.org/wonder/what-is-a-simple-machine*
Have more of your questions about simple machines answered on this fun website.

**Publisher's note to educators and parents:** Our editors have carefully reviewed these websites to ensure that they are suitable for students. Many websites change frequently, however, and we cannot guarantee that a site's future contents will continue to meet our high standards of quality and educational value. Be advised that students should be closely supervised whenever they access the internet.

# Index